YOUR KNOWLEDGE HAS VALUE

Michael Dienst

Nonorthodox Behavior of Fish Fins

Intelligent Mechanics (i-mech) in Nature and Design

GRIN Verlag

Bibliografische Information der Deutschen Nationalbibliothek:

Die Deutsche Bibliothek verzeichnet diese Publikation in der Deutschen National-
bibliografie; detaillierte bibliografische Daten sind im Internet über http://dnb.d-
nb.de/ abrufbar.

Imprint:

Copyright © 2013 GRIN Verlag GmbH
Druck und Bindung: Books on Demand GmbH, Norderstedt Germany
ISBN: 978-3-656-44320-9

This book at GRIN:

http://www.grin.com/en/e-book/215436/nonorthodox-behavior-of-fish-fins

GRIN - Your knowledge has value

Der GRIN Verlag publiziert seit 1998 wissenschaftliche Arbeiten von Studenten, Hochschullehrern und anderen Akademikern als eBook und gedrucktes Buch. Die Verlagswebsite www.grin.com ist die ideale Plattform zur Veröffentlichung von Hausarbeiten, Abschlussarbeiten, wissenschaftlichen Aufsätzen, Dissertationen und Fachbüchern.

Visit us on the internet:

http://www.grin.com/

http://www.facebook.com/grincom

http://www.twitter.com/grin_com

About the nonorthodox behavior of fish fins

Intelligent Mechanics (i-mech) in Nature and Design

Beuth University of Applied Sciences Berlin, Germany
Department VIII - Mechanical Engineering, Process
and Environmental Technology / Bionic Research Unit
Dipl.-Ing. Michael Dienst

Abstract. *The paper presents the state of the Systems Biology of rays in fin membranes of fish. The description of the functional geometry of the fin membrane is preceded by an anatomical taxis. Results of calculations and measurements for the mechanics of fin rays are identified.*

The aim of bionics is to decipher principles of living nature for to carry in technology and develop clever solutions [Rech-94][Bapp-99][Bech-97][Nach-98][Nach-00]. System Biology plays a crucial role: It produces the material of which the bionic generates innovations.

The ray-finned fishes are a very successful class of bony fish. To the genuine bony fish include, with approximately 30000 known extant species about 96 percent of all living fish species and about half of all vertebrate species described. Their anatonmie and the mechanics of their musculoskeletal system is the subject of numerous studies. Nevertheless, the considerable diversity of function and design of fin rays, their evolutionary history, the individual growth and differentiation during individual development is little understood.
Assignments of features and functions of the fin rays of different species with different skills and habits, such as hunting, escape, rooting or different swimming styles are still partially unknown. Consider the fish fin

in the context of fish body. Fin rays are part of the vertebrate skeleton, which forms a series of solid, articulated (skeletal) elements that are important in cooperation with the muscles for the movement. In light of the evolutionary development of vertebrates, the spinal column is older than any part of the postcranial skeleton. About the origin of the extremities of the fish there despite new fossils, improved methods, sophisti-cated phylogenetic analysis to date, no consensus is among evolutionary biologists [W-01] [W-02] [W-03] [W-04].

In the early 1880s several anatomists postulated that the original vertebrate continuous, paired side fins from the gills had to vent, so-called fin edges [Hild-01]. It was assumed that the fins of modern fishes represent segments of those originally continuous fin edges. Modern fin edges supporting aquatic life are, for example, eels and lampreys. The genome of the sea lamprey has been recently deciphered [W-10]. The fact that lampreys are vertebrates show their inner cartilaginous skeleton and the structure of their brain. Lampreys are the only survivors of an ancient evolutionary line of vertebrates represent.

The visible membrane of fin fish may have been originally supported only in the course of evolution of dermal scales in the skin covering them. The fins of more advanced bone fish were stabilized by a series of slender rays in the inner region. Basically, the rays of the cartilaginous fish are slender, not articulated and elastic and called Ceratotrichia. Fin rays of fish bones are wider geglie amended, proximal pairs, branched distally and ossified and are called Lepidotrichia, they are evolutionarily derived from scales described [W-06][W-06][Hild 01]. The caudal fin is used for propulsion force production, to stabilize the non-driven rectilinear locomotion and maneuvering. If the animal in its fluidic environment locates inhomogeneities, a velocity field or a suitable pressure gradient, the fish can use this to its own mobility.

James Liao of Harvard University in Cambridge (Massachusetts) built a tank for an underwater landscape and examined the swimming behavior of the animals, with affixed electrodes to the fins. They came to the conclusion that the fish moves in a zigzag of eddy to eddy and needed relatively low muscle strength for this type of locomotion [W-07].

The interaction and exchange interaction of transported in a flow vortices with a fin membrane is a fundamental phenomenon of vortex flow and inversion-flow and subject of analysis of active and passive control of vertebrate aquatic species.

The principles of vortex control are of great importance for the understanding of how fish swim and maneuver. By Gopalkrishnan et al (1994), Armed Lien et al (1996) and Anderson (1996), a harmonically oscillating airfoil in a afflicted with large vortices interact favorably when both the eddy size and the frequency of the harmonically oscillating profile fit the harmoniously flow. Fluid-structure interaction of flexible

bodies in vortical flows is the subject of extant research [Gopa-94][Read-02][Ande-99][alb-09] [Liao-06][Tria-02] [Floc 09][Stre-96].

Consider the momentum exchange with the medium of the wing membrane of the fish fin. The energy transfer at the fluid-structure interaction can be productive or generative. In a productive interaction the fin couples energy force from the flow into the structure. In a generative fluid-structure interaction, the fin membrane couples energy from the structure into the fluid. Production and generation can take place in a time-locally intertwined, complex overall process.

Unlike technology, where the energy and information exchange in powerwings can be clearly described and assigned, the biological wing constructions provides a complex, for feedback and adaptation enabled multifunctional systems. These are optimized and able to control their fluid environment, interact with them and condition them for their transport and mobility issues. Lauder sum up, that in a swirling flow transported vortices the fluid-structure interaction is generative with a fin membrane, when the timing of body motion is synchronal with the shape of a Karman vortex street. The fluid-structure interaction of a vertebra with a fin membrane is productive generative, when the timing of body motion is synchronal with the shape of a inverted Karman vortex street. Periodicity, frequency, phase and direction of the vortex flow has a significant impact on the quality of the fluid-structure interaction with the fin membrane. From the point of view of bionics, adaptive foils are a way of passive flow control. This makes a profound research required.

In several research projects at the Beuth University of Applied Sciences Berlin, since 2006 biologistic backgrounds "intelligent mechanics" we considered the fundamental solution for auto-adaptive profiles of biological fins. First technical intelligent kinematics were designed in 2005 [MIR 05], numerical approaches in 2008 [KRE -08], systems with fluid-structure interaction were investigated [Sie-10], [Sie-11] and patents on adaptive load components were developed [USP-12][DEP-11].

Numerical models of fluid-structure interaction only exist for selected conditions. As part of future research, projects will aim to a process chain to develop the solutions of body deformation (finite element method, FEM) and flow field (computational fluid dynamics CFD) in a common simulation approach under the special conditions of highly complex dynamic couples (Fluid Structure Interaction FSI). Simulation and calculation results represent the basis for the design of real flow components.

State of science. Fish fins are propulsion systems with high efficiency. They are able to form an optimum shape of the fins over an process cycle. Fins are the system boundary between the structural driving apparatus of the animal and the surrounding fluid. Their special shape of

the inner fin fish reacts passively and without cognitive control effort of the organism to various fluid conditions. This, the biosystem inherent "intelligent mechanics" is the subject of our research and contributes to the analysis of locomotion fluidic creatures.

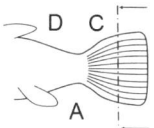

Figure 1
C caudal fin, anal and dorsal fin and D.

Fins made of a diaphragm support surface (webbing) which is stabilized by nonisotrop rays. In the muscles, the fin rays are anchored with fin ray beams. The architecture of this structure represents a balanced combination of stiffness and flexibility and allows the creature a finely tuned hydrodynamic interaction with its environment. The rays of the bone fish are still sting rays (hard) and link beams (soft) distinction. Hard jets are disjointed, mostly smooth pieces of bone, soft rays consist of two fused halve pipes.
In the soft rays is between (1) undivided, articulated and thorn-like and (2) split, divided, and (3) fan-like split, divided distinguished. The Figure 1 shows the caudal fin C, the anal and dorsal fins A and D.

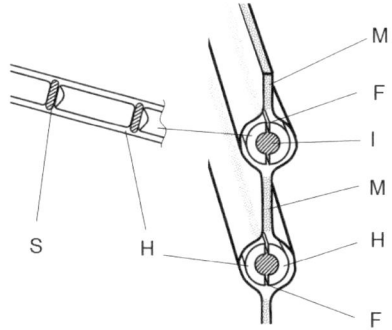

Figure 2.
Representation of a fins membrane.
Web S, halve pipe H, membrane M, Fugue F and Inlet I.

The soft rays of the caudal fin fish can imagine two at a certain distance associated with webs S, articulated half tubes H. The half-tubes system has an inlet I.   It fills the space between tubes and half webs. The membrane M coated tubes can slide on one another (Figure 2).
Fish fin rays are bilateral structures. The two halves of each beam to a certain extent slide on one each other. The sliding motion is in response external loads and / or when the bases of the fin system, the muscles are moved to the root of the fin rays. The surrounding membrane forms

bags, which encase the half tubes of fin rays and paste them into a compact quasi-round material and radially stabilize. The two half tubes are mechanically coupled together to the described tube system. Membrane with membrane bags and tubes systems, respectively, skin and fins fin rays form a three-dimensional wing with anisotropic properties.

Fish are able to actively control the bending of each fin to form the beam and the entire membrane in a very complex way. The decisive feature of the kinematic webbing membrane to perform a "non-orthodox" stress-strain interaction is based on the half-tube systems. Under horizontally to the main axis of the fish line load, the rays perform an elastic concave deformation. The curvature is opposite to the loading direction. Accustomed to associate such a simple beam load a convex curve, this concave stress-strain behavior seems paradoxical. The sketch, Figure 3 shows the fin of a (dead) Mackerel under point load.

When the momentum exchange at the membrane surface of caudal fin is very high, the biological control and control surface behaves flexible, resilient elastic and can yield a non-axial flow. The pressurization deformation interaction correlated with the direction of the force acting in the sense of a conventional stress-strain regime. Conventionally deformed components behave mechanically Orthodox. In normal operation, however, technically speaking the "design range of flow component" shows a fish fin mechanically nonorthodox, indeed paradoxical deformation behavior: a direction opposite to the force acting deformation realize paradoxical pressurization Deformation interactions.

Figure 3.
The non-Orthodox pressurization motion behavior of a mackerel fin.

Theoretical investigations and measurements. A calculation model of the curvature along the beam under load beam fins hydrodynamic (external) force has been proposed by McCutchen. The loads applied by the biological fin in the advancing operation can be a point force at any location along the fin beam from the root to fin-ray peak to a highly complex distributed load during the hydrodynamic interaction relate [MCCU-70].

In a seminal work of Alben, Madden and Lauder, building on the conclusions of McCutchen, the mechanical properties of fin rays are examined. The aim is to predict - on the basis of a linear elastic model - that deformation of the fin rays, resulting from muscular activity to the propulsive force production in swimming, to stabilize the non-driven rectilinear locomotion, for maneuvering and under the influence of any external forces.

The theoretical results are compared with the analysis of measurements of the force-displacement behavior of real rays. In the foreground are questions regarding the custom of the complex function of geometry in the interaction with the material properties of the rays. [Alb-06]. The finray model is based on the simplifying assumption of a symmetric equation of motion inextensible pair of beams under load (elastic theory). The bar is located between the thin layer of material has an influence on the forms, which may take the elastic fin rays. In the actual fin rays, the substance is inside a collagen gel web with complex yet experimentally verified properties [Antm 05] [BATC-67] [Sege-87].

The measurements are performed on a root fixed fin ray. Here, a position at the base of fins ray is selected in the full segmentation begins. The fin ray has at the root of a pre-steering, similar to an impressed by the fin muscle displacement force. This load case is the interests of Lauder. Is the displacement measured from the central (unloaded) central location with a photometric method. The geometric parameters are determined directly from the graphs. The rays have an approximately cylindrical geometry. Comparing the solution of theoretical model and actual results rays under a point load, it is quite significant differences in the self-adjusting curvature of the fin ray. Lauder may explain this discrepancy from the nonlinear elastic properties of the actual fin ray.

Conclusion. The level of knowledge about the anatomical design, the operational function and in particular about the pressurization motion behavior of the fish caudal fin is insufficient. A future research in the field of non-orthodox fluid-structure interaction with the target, according to the bionics to decipher biological phenomena and transfer in technology, should a multi-layered analysis of the principles of intelligent mechanics rays designing a reinforced membrane wings on the basis of experiments on real be preceded by system and with the support of computer models.

Bibliography for further reading, patents and web links.

[Albe-09]   Alben, S. (2009) On the swimming of a flexible body in a
              vortex street. in J. Fluid Mech. (2009), vol. 635, pp. 27–45.
              Cambridge University Press 2009
[Albe-06]   Alben, S., Madden, P.G., Lauder, V.L. (2006) The mechanics of active fin-
              shape control in ray finned fishes. Journal of the Royal Society. Interface Vol.:
              2007/4, S. 243-256.
[Ande-99]   Anderson, J.M. (1999) NEAR-BODY FLOW DYNAMICS IN SWIMMING FISH,
              The Journal of Experimental Biology 202, 2303–2327 (1999)
[Antm-05]   Antman S.S. (2005) Nonlinear problems of elasticity. 2nd edn. Springer; New
              York, NY.
[Batc-67]   Batchelor G.K. (1967) An introduction to fluid dynamics, 1st edn. Cambridge
              University Press; Cambridge, UK.
[Bann-02]   Bannasch, Rudolph. (2002) Vorbild Natur. In: design report 9/02, S.20ff.
              Blue.C Verlag Stuttgart.
[Bapp-99]   Bappert, R. Bionik, (1999) Zukunftstechnik lernt von der Natur.
              SiemensForum München/Berlin und Landesmuseum für Technik und Arbeit
              (Herausgeber).
[Barg-11]   Bagaric, B. (2011). Modellierung, Simulation und Parametrisierung eines
              virtuellen Strömungskanals mit dem Programmsystem FS-Flow.
              Untersuchung typischer Szenarien endlicher Traglügel. Bachelorarbeit a.d.
              BeuthHS Berlin (082011).
[Bech-93]   Bechert, D.W.: Verminderung des Strömungswiderstandes durch bionische
              Oberflächen. In: VDI-Technologieanalyse Bionik, S. 74 – 77. VDI-
              Technologiezentrum Düsseldorf 1993.
[Bech-97]   Bechert, D.W., Biological Surfaces and their Technological Application. 28th
              AIAA Fluid Dynamics Conference: 1997
[Curr-25]   Curry, M. (1925) Die Aerodynamik des Segels und die Kunst des Regatta-
              Segelns. Diessen vor München: Jos. C. Huber, 1925.
Floc-09]   France Floch,F. Laurens, J.M. (2009) Comparison of hydrodynamics
              performances of a porpoising foil and a propeller. in: First International
              Symposium on Marine Propulsors smp'09, Trondheim, Norway, June 2009
[Gopa-94]   Gopalkrishnan, R.(1994) Active vorticity control in a shear flow using a
              flapping foil. in J. Fluid Mech. (1994), vol. 274, pp. 1-21 Cambridge University
              Press.
[Hild-01]   Hildebrand, M., Goslow, G.E., (2001) Vergleichende und funktionelle
              Anatomie der Wirbeltiere. Springer Verlag Berlin, N.Y.
[Liao-03]   Liao, J.C.; Beal, D.; Lauder, G.; Triantayllou, M. (2003): Fish Exploting
              Vortices Decrease Muscle Activty, In: Science 2003, S. 1566-1569. AAAS.
[Liao-06]   Liao, J.C.; Passive propulsion in vortex wakes. in J. Fluid Mech. (2006), vol.
              549, pp. 385–402. c_ 2006 Cambridge University Press
[Kreb-08-2]   Krebber, B.: "i-mech". Untersuchung der intelligenten Mechanik von
              Fischflossen mit Hilfe von FSI- Simulation. Forschungsbericht der
              Technischen Fachhochschule Berlin 2007/08
[Kreb-08-1]   Krebber, B., H.-D. Kleinschrodt und K. Hochkirch: (2008) Fluid-Struktur-
              Simulation zur Untersuchung intelligenter Mechanik von Fischflossen. ANSYS
              Conference & 26. CADFEM Users´ Meeting,
[McCu-70]   McCutchen C.W. (1970) The trout tail fin, a self-cambering hydrofoil. J.
              Biomech. 1970/3, S. 271–281.
[Nach-98]   Nachtigall, W.: Bionik. Grundlagen und Beispiele für Ingenieure und
              Naturwissenschaftler. Springer-Verlag, Berlin-Heidelberg-New York 1998.
[Nach-00]   Nachtigall, W.; Blüchel, K. Das große Buch der Bionik. Stuttgart: Deutsche
              Verlags Anstalt: 2000.

[Mirs-05]	Mirtsch, F.; Dienst, M.: FlowBow-Artifizielle adaptive Strömungskörper nach dem Vorbild der Natur. In: Forschungsbericht der Technischen Fachhochschule Berlin 2005
[PaBe-93]	Pahl. G.; Beitz, W.: Konstruktionslehre, 3.Auflage. Berlin- Heidelberg- New York-London-Paris-Tokio: Springer 1993
[Pfeif-07]	Pfeiffer,Rolf; Bongard, Josh (2007): How the body shapes the way we think, The MIT Press
[Read-02]	D.A. Read (2002) Forces on oscillating foils for propulsion and maneuvering, in Journal of Fluids and Structures 17 (2003) 163–183 Cambridge University Press
[Rech-94]	Rechenberg, Ingo, (1994) Evolutionsstrategie. Frommann Holzboog Verlag Stuttgart- Bad Cannstatt.
[Sege-87]	Segel L.A. (1987) Mathematics applied to continuum mechanics. 1st edn. Dover Publications; New York, NY.
[Siew-10]	Siewert, M; Kleinschrodt, H-D; Krebber, B; Dienst, Mi. (2010) FSI- Analyse auto-adaptiver Profile für Strömungsleitflächen. In: Tagungsband, ANSYS Conference & 28th CADFEM Users' Meeting Aachen 2010.
[Siew-11]	Siewert, M; Kleinschrodt, H-D.(2011) Bionical Morphological Computation. In: Nachhaltige Forschung in Wachstumsbereichen Bd.1, Logos Verlag Berlin.
[Stre-96]	Streitlien, K. (1996) Efficient foil propulsion through vortex control, Aiaa Journal - AIAA J , vol. 34, no. 11, pp. 2315-2319, 1996
[Tria-95]	Triantafyllou, M. (1995): Effizienter Flossenantrieb für Schwimmroboter, Spektrum der Wissenschaft 08-1995, S. 66–73, Wiss. Verlagsges. mbH, Heidelberg 1995.
[Tria-02]	Triantafyllou, M. (2002) Vorticity Control in Fish-like Propulsion and Maneuvering, INTEGR. COMP. BIOL., 42:1026–1031 (2002)
[USP-12]	US Patent 13517181, Components Designed to be Loadadaptive, Dienst (2012).
[DEP-11]	DE Patent 2010/075164, Belastungsadaptiv ausgebildete Bauteile, Dienst (2011).
[USP-08]	US Patent US20110281479. Flexible impact blade with drive device for a flexible impact blade. Kniese, L. Bannasch, R. (2008).
[USP-13]	US Patent US20100263803. Door Element. Kniese, L. Bannasch, R. (2008).
[USP-10]	US Patent US8333417 B2. Manipulator tool and holding and/or expanding tool with at least one manipulator tool, . Kniese, L., Bannasch, R. (2010).

[W-01]	http://de.wikipedia.org/wiki/Strahlenflosser.(abgerufen27052013)
[W-02]	http://en.wikipedia.org/wiki/Osteichthyes.(abgerufen27052013)
[W-03]	http://www.pflegewiki.de/wiki/Cranial.(abgerufen27052013)
[W-04]	http://de.wikipedia.org/wiki/H%C3%A4malbogen. (abgerufen27052013)
[W-06]	http://fishbase.mnhn.fr/glossary/Glossary.php?q=ceratotrichia (abgerufen 27052013)
[W-07]	http://en.wiktionary.org/wiki/lepidotrichia (abgerufen 27052013)
[w-08]	http://sciencev1.orf.at/news/97597.html, Fische kreuzen energiesparend gegen den Strom. (abgerufen 27052013)
[W-10]	http://www.spiegel.de/wissenschaft/natur/neunauge-genom-entschluesselt-a-884712.html (abgerufen 28052013)

photographs

Figure 1.
Caudal fin C, anal and dorsal fin A and D.
Sketch Mi. Dienst (2013)

Figure 2.
Schematic representation of a fins membrane. Web S, halve tube H, membrane M, Fugue F and Inlet I.
Sketch Mi. Dienst (2013)

Figure 3.
The non-Orthodox pressurization motion behavior of a mackerel fin.
Photographic representation, Mi. Dienst (2008)

Contact:

Dipl.-Ing. Michael Dienst
Beuth University of Applied Sciences Berlin, Germany
Department VIII - Mechanical Engineering, Process and Environmental Technology.
Bionic Research Unit
Luxemburger Str. 10,
D - 13353 Berlin-Wedding

MiDienst@beuth-hochschule.de
http:// www.beuth-hochschule.de

The BIONIC RESEARCH UNIT is a research-based professional group of teachers and students at the Beuth University of Applied Sciences Berlin and partner for industrial services in the knowledge field of bionics.

http://projekt.beuth-hochschule.de/bru